This Book belongs

To

....

....

....

MONTH

Date	Location / Details	Price Per	Amount
		Total	

MONTH

Date	Location / Details	Price Per	Amount
		Total	

MONTH

Date	Location / Details	Price Per	Amount
		Total	

MONTH

Date	Location / Details	Price Per	Amount

Total

MONTH

Date	Location / Details	Price Per	Amount
		Total	

MONTH

Date	Location / Details	Price Per	Amount
		Total	

MONTH

Date	Location / Details	Price Per	Amount

Total

MONTH

Date	Location / Details	Price Per	Amount
		Total	

MONTH

Date	Location / Details	Price Per	Amount
		Total	

MONTH

Date	Location / Details	Price Per	Amount

Total

MONTH

Date	Location / Details	Price Per	Amount
		Total	

MONTH

Date	Location / Details	Price Per	Amount

Total

MONTH

Date	Location / Details	Price Per	Amount
		Total	

MONTH

Date	Location / Details	Price Per	Amount
		Total	

MONTH

Date	Location / Details	Price Per	Amount

Total

MONTH

Date	Location / Details	Price Per	Amount

Total

MONTH

Date	Location / Details	Price Per	Amount

Total

MONTH

Date	Location / Details	Price Per	Amount

Total

MONTH

Date	Location / Details	Price Per	Amount
		Total	

MONTH

Date	Location / Details	Price Per	Amount
		Total	

MONTH

Date	Location / Details	Price Per	Amount

Total

MONTH

Date	Location / Details	Price Per	Amount
		Total	

MONTH

Date	Location / Details	Price Per	Amount
		Total	

MONTH

Date	Location / Details	Price Per	Amount
		Total	

MONTH

Date	Location / Details	Price Per	Amount

Total

MONTH

Date	Location / Details	Price Per	Amount
		Total	

MONTH

Date	Location / Details	Price Per	Amount

Total

MONTH

Date	Location / Details	Price Per	Amount
		Total	

MONTH

Date	Location / Details	Price Per	Amount
		Total	

MONTH

Date	Location / Details	Price Per	Amount
		Total	

MONTH

Date	Location / Details	Price Per	Amount
		Total	

MONTH

Date	Location / Details	Price Per	Amount
		Total	

MONTH

Date	Location / Details	Price Per	Amount

Total

MONTH

Date	Location / Details	Price Per	Amount

Total

MONTH

Date	Location / Details	Price Per	Amount
		Total	

MONTH

Date	Location / Details	Price Per	Amount
		Total	

MONTH

Date	Location / Details	Price Per	Amount

Total

MONTH

Date	Location / Details	Price Per	Amount

Total

MONTH

Date	Location / Details	Price Per	Amount

Total

MONTH

Date	Location / Details	Price Per	Amount
		Total	

MONTH

Date	Location / Details	Price Per	Amount

Total

MONTH

Date	Location / Details	Price Per	Amount

Total

MONTH

Date	Location / Details	Price Per	Amount

Total

MONTH

Date	Location / Details	Price Per	Amount

Total

MONTH

Date	Location / Details	Price Per	Amount
		Total	

MONTH

Date	Location / Details	Price Per	Amount
		Total	

MONTH

Date	Location / Details	Price Per	Amount

Total

MONTH

Date	Location / Details	Price Per	Amount

Total

MONTH

Date	Location / Details	Price Per	Amount

Total

MONTH

Date	Location / Details	Price Per	Amount

Total

MONTH

Date	Location / Details	Price Per	Amount

Total

MONTH

Date	Location / Details	Price Per	Amount
		Total	

MONTH

Date	Location / Details	Price Per	Amount
		Total	

MONTH

Date	Location / Details	Price Per	Amount
		Total	

MONTH

Date	Location / Details	Price Per	Amount

Total

MONTH

Date	Location / Details	Price Per	Amount

Total

MONTH

Date	Location / Details	Price Per	Amount
		Total	

MONTH

Date	Location / Details	Price Per	Amount

Total

MONTH

Date	Location / Details	Price Per	Amount
		Total	

MONTH

Date	Location / Details	Price Per	Amount

Total

MONTH

Date	Location / Details	Price Per	Amount

Total

MONTH

Date	Location / Details	Price Per	Amount
		Total	

MONTH

Date	Location / Details	Price Per	Amount
		Total	

MONTH

Date	Location / Details	Price Per	Amount
		Total	

MONTH

Date	Location / Details	Price Per	Amount

Total

MONTH

Date	Location / Details	Price Per	Amount

Total

MONTH

Date	Location / Details	Price Per	Amount
		Total	

MONTH

Date	Location / Details	Price Per	Amount

Total

MONTH

Date	Location / Details	Price Per	Amount
		Total	

MONTH

Date	Location / Details	Price Per	Amount

Total

MONTH

Date	Location / Details	Price Per	Amount
		Total	

MONTH

Date	Location / Details	Price Per	Amount

Total

MONTH

Date	Location / Details	Price Per	Amount

Total

MONTH

Date	Location / Details	Price Per	Amount
		Total	

MONTH

Date	Location / Details	Price Per	Amount

Total

MONTH

Date	Location / Details	Price Per	Amount
		Total	

MONTH

Date	Location / Details	Price Per	Amount
		Total	

MONTH

Date	Location / Details	Price Per	Amount
		Total	

MONTH

Date	Location / Details	Price Per	Amount
		Total	

MONTH

Date	Location / Details	Price Per	Amount
		Total	

MONTH

Date	Location / Details	Price Per	Amount
		Total	

MONTH

Date	Location / Details	Price Per	Amount

Total

MONTH

Date	Location / Details	Price Per	Amount

Total

MONTH

Date	Location / Details	Price Per	Amount
		Total	

MONTH

Date	Location / Details	Price Per	Amount
		Total	

MONTH

Date	Location / Details	Price Per	Amount

Total

MONTH

Date	Location / Details	Price Per	Amount

Total

MONTH

Date	Location / Details	Price Per	Amount
		Total	

MONTH

Date	Location / Details	Price Per	Amount
		Total	

MONTH

Date	Location / Details	Price Per	Amount

Total

MONTH

Date	Location / Details	Price Per	Amount
		Total	

MONTH

Date	Location / Details	Price Per	Amount
		Total	

MONTH

Date	Location / Details	Price Per	Amount
		Total	

MONTH

Date	Location / Details	Price Per	Amount
		Total	

MONTH

Date	Location / Details	Price Per	Amount

Total

MONTH

Date	Location / Details	Price Per	Amount
		Total	

MONTH

Date	Location / Details	Price Per	Amount
		Total	

MONTH

Date	Location / Details	Price Per	Amount
		Total	

MONTH

Date	Location / Details	Price Per	Amount

Total

CPSIA information can be obtained
at www.ICGtesting.com
Printed in the USA
LVHW110333291019
635549LV00011B/4704/P

9 781700 052001